Animal Consciousness

Bridging Worlds: Exploring the Depths of Animal Consciousness

Bienuel Tenio

Animal Consciousness
Bridging Worlds: Exploring the Depths of Animal Consciousness

Copyright © 2024 by Bienuel Tenio

All rights reserved. No part of this publication may be reproduced, distributed, or transmitted in any form or by any means, including photocopying, recording, or other electronic or mechanical methods, without the prior written permission of the publisher, except in the case of brief quotations embodied in critical reviews and certain other noncommercial uses permitted by copyright law.

For permissions requests, write to the publisher at the address below.

Kindle Direct Publishing
410 Terry Avenue North
Seattle, WA 98109-5210
USA

Preface

In the intricate tapestry of existence, one thread remains profoundly woven into the fabric of life itself: consciousness. From the depths of the oceans to the sprawling savannas, from the highest peaks to the hidden depths of the forest floor, consciousness manifests in myriad forms, transcending species boundaries and captivating our imagination.

Welcome to "Animal Consciousness / Bridging Worlds: Exploring the Depths of Animal Consciousness." Within these pages, we embark on a journey that transcends the familiar landscapes of human understanding, delving into the rich and diverse realms of animal consciousness.

In our quest for enlightenment, we begin at the genesis – Chapter 1: Introduction. Here, we set the stage, recognizing the profound importance of understanding animal consciousness. As we navigate through the corridors of inquiry, we endeavor to define consciousness itself, daring to explore its enigmatic essence across species. Through this exploration, we illuminate the path ahead, offering a guiding beacon for the journey that lies before us.

Chapter 2: The Evolution of Consciousness invites us to embark on a voyage through time, tracing the origins of consciousness as it unfolds across the vast expanse of the animal kingdom. Drawing upon the insights gleaned from comparative neurology and evolutionary theory, we unravel the intricate tapestry of consciousness's evolution – from its humble beginnings to the pinnacle of self-reflection.

As we venture deeper into the labyrinth of understanding, Chapter 3: Perception and Sensory Experience beckons us to explore the sensory worlds of our fellow inhabitants. Through their eyes, ears, and senses beyond our comprehension, we gain a newfound appreciation for the diversity of conscious experience that permeates the natural world.

In Chapter 4: Emotions and Social Intelligence, we delve into the heart of sentient existence, exploring the kaleidoscope of emotions that color the lives of animals. From joy to fear, from empathy to conflict resolution, we bear witness to the rich tapestry of emotional complexity that binds us all.

Language, Communication, and Symbolism emerge as the focal point of Chapter 5, inviting us to transcend the boundaries of linguistic convention and delve into the realm of nonverbal and symbolic communication. Through extraordinary case studies and the unraveling of ancient mysteries, we seek to decipher the languages that echo through the corridors of animal consciousness.

Chapter 6: Consciousness and Intelligence beckons us to contemplate the mysteries of cognition and problem-solving prowess that define the boundaries of animal intelligence. Through tales of innovation and cognitive flexibility, we confront the enigma of animal minds, challenging the very foundations of anthropocentric perspectives.

In Chapter 7: Consciousness in the Wild, we venture into the untamed wilderness, bearing witness to the raw beauty of animal consciousness as it unfolds in its natural habitat. From field studies to the ethical considerations of captive settings, we uncover the hidden truths that lie beneath the surface of our understanding.

The Human-Animal Connection serves as the focal point of our exploration in Chapter 8, inviting us to forge bonds of empathy and understanding with our fellow inhabitants. Through historical perspectives and shared experiences, we glimpse the profound interconnectedness that binds us to the tapestry of life itself.

As we embark on this odyssey of exploration, let us remember that our journey is but a fleeting moment in the vast expanse of time. May the pages of this book serve as a testament to the enduring beauty of animal consciousness and the boundless depths of our shared existence.

Together, let us bridge worlds and explore the infinite depths of animal consciousness that lie beyond the confines of human comprehension.

Table of Contents

Chapter 1: Introduction 9

   I. Setting the Stage: The Importance of Understanding Animal Consciousness 9

   II. Defining Consciousness: Exploring the Concept Across Species 10

   III. Overview of the Book's Structure 11

Chapter 2: The Evolution of Consciousness 13

   I. Origins of Consciousness: Tracing Its Roots in the Animal Kingdom. 13

   II. Comparative Neurology: Insights from Brain Anatomy and Function. 14

   III. Theories of Consciousness Evolution: From Simple Awareness to Self-Reflection 15

Chapter 3: Perception and Sensory Experience 18

   I. Sensory Worlds: How Animals Perceive Their Environments. 18

   II. The Role of Senses in Conscious Experience. 19

   III. Cross-Species Comparisons: Understanding Unique Sensory Perceptions. 20

Chapter 4: Emotions and Social Intelligence 22

   I. Emotional Lives of Animals: Exploring Joy, Fear, and Empathy. 22

   II. Social Dynamics: Communication, Cooperation, and Conflict Resolution. 23

   III. Ethical Considerations: Implications for Animal Welfare and Rights. 24

Chapter 5: Language, Communication, and Symbolism 26

I. Communication Beyond Words: Examining Nonverbal and Symbolic Language. 26

II. Case Studies: Extraordinary Examples of Animal Communication. 27

III. Exploring the Limits of Human Understanding: Deciphering Animal Languages. 28

Chapter 6: Consciousness and Intelligence 30

I. Problem-Solving Abilities: Insights into Animal Intelligence. 30

II. Tool Use and Innovation: Demonstrations of Cognitive Flexibility. 31

III. The Enigma of Animal Minds: Challenging Anthropocentric Perspectives. 32

Chapter 7: Consciousness in the Wild: Observations and Discoveries 34

I. Field Studies: Observing Animal Behavior in Natural Habitats. 34

II. Insights from Captive Settings: Ethical Considerations and Research Challenges. 35

III. Revelations from Unexpected Sources: Surprising Discoveries in Animal Consciousness. 36

Chapter 8: The Human-Animal Connection 38

I. Companion Animals: Bonds, Empathy, and Understanding. 38

II. Historical Perspectives: Cultural Views on Animal Consciousness. 39

III. Shared Experiences: Learning from Animals and Enriching Human Lives. 40

Chapter 9: Consciousness and Conservation 42

I. Biodiversity and Ecosystem Health: Recognizing Animals as Sentient Beings. 42

II. Conservation Ethics: Protecting Animal Consciousness in the Wild. 43

III. Future Directions: Integrating Consciousness Studies into Conservation Efforts. 44

Chapter 10: Beyond Boundaries: Philosophical and Metaphysical Reflections 46

I. The Nature of Consciousness: Exploring Mysteries and Paradoxes. 46

II. Interconnectedness and Interdependence: Rethinking Humanity's Relationship with Animals. 47

III. Towards a Holistic Understanding: Integrating Science, Ethics, and Spirituality. 48

Chapter 11: Conclusion 50

I. Summarizing Key Insights and Findings. 50

II. Looking Ahead: Future Directions in Animal Consciousness Research. 51

III. Call to Action: Embracing a World Where All Conscious Beings Matter. 52

Chapter 1: Introduction

I. Setting the Stage: The Importance of Understanding Animal Consciousness

In the vast landscape of philosophical inquiry, few questions are as fundamental and compelling as the exploration of consciousness. At the heart of this inquiry lies a profound curiosity about the nature of existence and the boundaries of awareness. In "Animal Consciousness / Bridging Worlds: Exploring the Depths of Animal Consciousness," Bienuel Tenio embarks on a journey to illuminate the intricate tapestry of animal consciousness, inviting readers to transcend the limitations of human-centric perspectives and embrace a more inclusive understanding of sentient beings.

The importance of understanding animal consciousness extends far beyond the realm of academic inquiry. It strikes at the core of our ethical responsibilities and moral considerations toward the diverse array of species that inhabit our planet. As stewards of the Earth, we are entrusted with the care and preservation of all living beings, each endowed with its own unique capacity for experience and awareness. By delving into the depths of animal consciousness, we not only enrich our understanding of the natural world but also cultivate empathy, compassion, and reverence for all forms of life.

Moreover, the exploration of animal consciousness holds profound implications for fields ranging from neuroscience and psychology to ethics and environmental conservation. By unraveling the mysteries of animal cognition and perception, we gain invaluable insights into the richness and complexity of non-human minds, challenging prevailing assumptions and fostering a deeper appreciation for the interconnectedness of all living beings.

II. Defining Consciousness: Exploring the Concept Across Species

At the heart of the inquiry into animal consciousness lies the elusive concept of consciousness itself—a topic that has captivated philosophers, scientists, and thinkers across millennia. As we embark on our exploration, we are confronted with a myriad of questions: What defines consciousness? How do we recognize its presence in different species? Can consciousness exist beyond the confines of human experience?

In "Animal Consciousness / Bridging Worlds," Bienuel Tenio invites readers to embark on a nuanced exploration of consciousness that transcends traditional boundaries and embraces the diversity of sentient experience. Drawing upon insights from neuroscience, psychology, ethology, and philosophy, Tenio challenges us to reconsider our preconceived notions of consciousness and embrace a more inclusive framework that acknowledges the varied manifestations of awareness across species.

Through a comparative approach, Tenio illuminates the continuum of consciousness that spans the animal kingdom, from the simplest organisms to the most complex vertebrates. By examining the behavioral, cognitive, and neural correlates of consciousness across diverse taxa, we gain a deeper appreciation for the richness and diversity of conscious experience, transcending anthropocentric biases and embracing the unity of life.

III. Overview of the Book's Structure

"Animal Consciousness / Bridging Worlds" is structured as a comprehensive exploration of animal consciousness, spanning multiple disciplines and perspectives. Through a series of thematic chapters, Tenio guides readers on a journey through the depths of sentient experience, inviting us to contemplate the mysteries of consciousness and the intricacies of interspecies communication.

The book is divided into three main sections, each offering a distinct vantage point from which to explore the complexities of animal consciousness. In the first section, "Foundations of Consciousness," Tenio lays the groundwork for our inquiry, delving into the historical, philosophical, and scientific dimensions of consciousness studies. Through a nuanced examination of key concepts and theories, we gain a deeper understanding of the foundations upon which our exploration is built.

In the second section, "Exploring Animal Minds," Tenio invites readers to embark on a journey through the diverse landscapes of animal cognition, perception, and emotion. Through a series of case studies and empirical investigations, we encounter the remarkable capacities of non-human animals and confront the ethical and philosophical implications of our findings.

Finally, in the third section, "Bridging Worlds," Tenio charts a course toward integration and synthesis, weaving together insights from disparate disciplines to forge a more holistic understanding of animal consciousness. Through interdisciplinary dialogue and collaborative inquiry, we strive to bridge the divide between human and non-human minds, embracing a worldview that honors the interconnectedness of all living beings.

In conclusion, "Animal Consciousness / Bridging Worlds" stands as a testament to the power of inquiry, imagination, and empathy in our quest to unravel the mysteries of consciousness and embrace the diversity of sentient experience. Through rigorous scholarship, compassionate inquiry, and visionary exploration, Bienuel Tenio invites us to embark on a transformative journey—one that transcends boundaries, challenges assumptions, and illuminates the profound depths of animal consciousness.

Chapter 2: The Evolution of Consciousness

I. Origins of Consciousness: Tracing Its Roots in the Animal Kingdom.

The journey to understand the evolution of consciousness is a profound exploration that takes us back to the origins of life itself. In "Animal Consciousness / Bridging Worlds," Bienuel Tenio embarks on a thoughtful examination of this journey, inviting readers to trace the intricate tapestry of consciousness through the annals of evolutionary history.

At the heart of our inquiry lies a fundamental question: Where did consciousness originate, and how did it evolve across the vast expanse of geological time? The quest for answers takes us deep into the heart of the animal kingdom, where we encounter a diverse array of sentient beings, each endowed with its own unique capacity for experience and awareness.

Through a meticulous examination of evolutionary evidence, Tenio illuminates the gradual emergence of consciousness from its primordial origins. From the simplest single-celled organisms to the complex neural networks of vertebrates, we witness the incremental steps that gave rise to the profound depths of consciousness we observe in modern-day organisms.

The origins of consciousness are rooted in the adaptive strategies of early life forms, where sensory perception, response to stimuli, and behavioral plasticity laid the groundwork for more complex forms of awareness. As we traverse the evolutionary timeline, we encounter the emergence of rudimentary forms of consciousness in primitive organisms, marked by the capacity for sensation, perception, and basic forms of cognition.

Yet, consciousness is not a static phenomenon confined to a single point in time or space. Rather, it is a dynamic and ever-evolving process that continues to shape the trajectory of life on Earth. By tracing its roots in the animal kingdom, we gain a deeper appreciation for the interconnectedness of all living beings and the profound continuity of consciousness across diverse taxa.

II. Comparative Neurology: Insights from Brain Anatomy and Function.

The study of comparative neurology offers invaluable insights into the evolution of consciousness, providing a window into the intricate workings of the brain across different species. In "Animal Consciousness / Bridging Worlds," Bienuel Tenio delves into this fascinating realm of inquiry, unraveling the mysteries of brain anatomy and function to shed light on the evolution of consciousness.

At the core of comparative neurology lies the recognition that the brain serves as the seat of consciousness—a complex network of neurons, synapses, and neural circuits that give rise to the rich tapestry of subjective experience. By examining the structural and functional similarities and differences across species, we gain a deeper understanding of the neural substrates underlying consciousness.

Through meticulous comparative analyses, Tenio reveals the remarkable diversity of neural architectures that have evolved in response to the unique ecological challenges and adaptive pressures faced by different organisms. From the intricate convolutions of the human cerebral cortex to the streamlined neural pathways of invertebrate taxa, each brain offers a window into the adaptive strategies and cognitive capacities of its respective bearer.

The insights gleaned from comparative neurology challenge traditional anthropocentric views of consciousness, highlighting the continuum of cognitive abilities that spans the animal kingdom. While the human brain may boast unparalleled complexity and computational power, it is but one node in the vast network of sentient minds that populate our planet.

III. Theories of Consciousness Evolution: From Simple Awareness to Self-Reflection

The evolution of consciousness has long been a subject of speculation and debate among philosophers, scientists, and thinkers across disciplines. In "Animal Consciousness / Bridging Worlds," Bienuel Tenio navigates this intellectual landscape with precision and insight, exploring the diverse theories that seek to elucidate the trajectory of consciousness evolution.

From the earliest stirrings of awareness in primitive organisms to the sophisticated self-reflection observed in certain higher vertebrates, the journey of consciousness is marked by a tapestry of transitions and transformations. At the heart of our inquiry lies the quest to unravel the underlying principles and mechanisms that have guided this remarkable process of emergence and elaboration.

One prominent theory posits that consciousness arose as an adaptive response to the complexities of the environment, enabling organisms to navigate and interact with their surroundings in more nuanced and adaptive ways. From this perspective, consciousness is viewed as a dynamic and evolving phenomenon shaped by the interplay of genetic predispositions, environmental stimuli, and evolutionary pressures.

Another theory suggests that consciousness emerged as a byproduct of neural complexity, arising from the intricate interconnections and feedback loops that characterize the brain. From this vantage point, consciousness is seen as an emergent property of neural dynamics—a phenomenon that transcends the sum of its individual parts and manifests as subjective experience.

Yet, perhaps the most intriguing aspect of consciousness evolution lies in its potential for self-reflection and self-awareness. In certain species, we observe the emergence of introspective capacities—the ability to reflect upon one's own thoughts, emotions, and experiences with a degree of metacognitive awareness. This capacity for self-reflection represents a profound leap in the evolution of consciousness, opening new vistas of inquiry into the nature of subjective experience and the boundaries of selfhood.

In conclusion, the evolution of consciousness is a multifaceted and enigmatic phenomenon that defies easy categorization or explanation. In "Animal Consciousness / Bridging Worlds," Bienuel Tenio invites readers to embark on a transformative journey—one that transcends disciplinary boundaries, challenges conventional wisdom, and illuminates the profound depths of sentient experience. Through rigorous scholarship, compassionate inquiry, and visionary exploration, we strive to unravel the mysteries of consciousness and embrace the diversity of life that surrounds us.

Chapter 3: Perception and Sensory Experience

I. Sensory Worlds: How Animals Perceive Their Environments.

In the tapestry of consciousness, perception serves as the gateway through which sentient beings interact with and make sense of their environments. In "Animal Consciousness / Bridging Worlds," Bienuel Tenio delves into the intricate realm of sensory experience, inviting readers to explore the rich and diverse worlds that unfold through the senses of animals.

At the heart of sensory perception lies a profound interplay between biology, ecology, and behavior. From the whispering whispers of wind to the kaleidoscope of colors that paint the world, each sensory modality offers a unique window into the rich tapestry of existence. Through a symphony of sights, sounds, smells, tastes, and touches, animals navigate their environments, forge social bonds, and perceive the subtle nuances of their surroundings.

The sensory worlds of animals are as varied as the ecosystems they inhabit, shaped by evolutionary pressures, ecological niches, and adaptive strategies. From the sonorous songs of whales that reverberate through the depths of the ocean to the intricate dances of fireflies that illuminate the night sky, each species perceives the world through a lens uniquely attuned to its needs and experiences.

In the exploration of sensory perception, we are called to transcend human-centric perspectives and embrace the diversity of sensory modalities that populate the natural world. Through the eyes, ears, noses, tongues, and bodies of animals, we gain a deeper appreciation for the richness and complexity of conscious experience, transcending anthropocentric biases and embracing the unity of life.

II. The Role of Senses in Conscious Experience.

Central to the fabric of consciousness lies the intricate interplay between sensory perception and conscious experience. In "Animal Consciousness / Bridging Worlds," Bienuel Tenio illuminates the profound role that senses play in shaping the contours of subjective reality, inviting readers to contemplate the intricate dance between sensation and cognition.

At its essence, consciousness emerges as a tapestry woven from the threads of sensory experience—a mosaic of sights, sounds, smells, tastes, and touches that converge to form the rich tapestry of subjective reality. Through the lens of sensory perception, animals engage with the world, discerning patterns, detecting threats, and forging connections that define their existence.

Yet, the role of senses in consciousness extends beyond mere perception, encompassing a dynamic interplay between sensation, attention, memory, and emotion. From the haunting call of a distant predator that sends shivers down the spine to the tantalizing aroma of ripe fruit that evokes memories of past pleasures, sensory experiences shape the contours of conscious awareness, imbuing life with meaning, depth, and significance.

In the exploration of consciousness, we are called to recognize the profound role that senses play in shaping our understanding of self and other. Through the kaleidoscope of sensory experience, we glimpse the rich tapestry of subjective reality, transcending the boundaries of human perception and embracing the diversity of conscious experience that animates the natural world.

III. Cross-Species Comparisons: Understanding Unique Sensory Perceptions.

As we journey deeper into the heart of consciousness, we are confronted with the remarkable diversity of sensory perceptions that populate the animal kingdom. In "Animal Consciousness / Bridging Worlds," Bienuel Tenio invites readers to embark on a comparative exploration of sensory modalities, illuminating the unique ways in which different species perceive and interact with their environments.

Across the vast expanse of evolutionary time, animals have evolved an extraordinary array of sensory adaptations, each finely tuned to the demands of their ecological niche. From the acute olfactory senses of dogs that discern the faintest scent trails to the ultrasonic echolocation of bats that pierce the darkness with pinpoint accuracy, each species perceives the world through a lens uniquely shaped by its evolutionary history and ecological context.

In the exploration of cross-species comparisons, we are called to transcend human-centric perspectives and embrace the richness and diversity of sensory perceptions that populate the natural world. Through the eyes of birds, the ears of dolphins, and the noses of insects, we gain a deeper appreciation for the kaleidoscope of conscious experience that animates the tapestry of life.

In conclusion, "Animal Consciousness / Bridging Worlds" stands as a testament to the power of sensory perception in shaping the contours of conscious experience. Through a thoughtful exploration of sensory worlds, the role of senses in consciousness, and cross-species comparisons, Bienuel Tenio invites us to transcend the limitations of human perception and embrace the diversity of sensory experiences that unite us with the vibrant tapestry of life.

Chapter 4: Emotions and Social Intelligence

I. Emotional Lives of Animals: Exploring Joy, Fear, and Empathy.

In the intricate tapestry of consciousness, emotions emerge as the vibrant threads that weave together the fabric of sentient experience. In "Animal Consciousness / Bridging Worlds," Bienuel Tenio invites readers to embark on a profound exploration of the emotional lives of animals, delving into the rich tapestry of joy, fear, and empathy that animates the natural world.

At the heart of emotional experience lies a profound depth of feeling—a kaleidoscope of sensations, desires, and intentions that shape the contours of conscious awareness. From the exuberant playfulness of young animals to the mournful cries of mourning elephants, each species expresses emotions in its own unique language, reflecting the rich diversity of conscious experience that permeates the natural world.

The exploration of animal emotions challenges traditional notions of human exceptionalism, inviting us to recognize the complex inner lives of non-human beings. Through a symphony of behaviors, vocalizations, and physiological responses, animals convey a profound depth of feeling, transcending the boundaries of language and culture to touch the hearts and minds of those who encounter them.

In the exploration of joy, fear, and empathy, we are called to transcend human-centric perspectives and embrace the universality of emotional experience that binds us with the web of life. Through the eyes of animals, we glimpse the raw beauty and vulnerability of conscious awareness, forging connections that transcend species boundaries and unite us in a shared tapestry of existence.

II. Social Dynamics: Communication, Cooperation, and Conflict Resolution.

Central to the fabric of consciousness lies the intricate dance of social dynamics—a symphony of communication, cooperation, and conflict resolution that shapes the tapestry of social life. In "Animal Consciousness / Bridging Worlds," Bienuel Tenio illuminates the profound role of social intelligence in shaping the contours of conscious experience, inviting readers to explore the rich tapestry of social interactions that animate the natural world.

At its essence, social intelligence emerges as a testament to the power of cooperation and collaboration—a dynamic interplay between individuals that transcends the boundaries of self and other. From the intricate courtship rituals of birds to the elaborate hierarchies of primate societies, animals navigate complex social landscapes, forging bonds that define the fabric of their communities.

Through a symphony of vocalizations, gestures, and behaviors, animals communicate a rich tapestry of meaning, transcending the limitations of language to convey emotions, intentions, and desires. From the rhythmic dance of honeybees that signals the location of nectar to the haunting calls of whales that reverberate through the depths of the ocean, each species possesses its own unique lexicon of communication that shapes the contours of social life.

In the exploration of social dynamics, we are called to recognize the profound interconnectedness of all living beings—a web of relationships that spans species, ecosystems, and generations. Through the lens of social intelligence, we glimpse the raw beauty and complexity of conscious awareness, forging connections that transcend the boundaries of self and other to embrace the unity of life.

III. Ethical Considerations: Implications for Animal Welfare and Rights.

As we navigate the depths of animal consciousness, we are confronted with profound ethical considerations that touch upon the very fabric of our moral landscape. In "Animal Consciousness / Bridging Worlds," Bienuel Tenio challenges us to confront the ethical implications of our interactions with non-human beings, inviting readers to reflect on our responsibilities toward the sentient creatures that share our world.

At the heart of ethical inquiry lies a fundamental recognition of the intrinsic value and dignity of all living beings—a recognition that transcends species boundaries and embraces the interconnectedness of life. From the ethical imperative to alleviate suffering to the moral obligation to respect the autonomy and agency of sentient beings, our ethical responsibilities toward animals compel us to confront the complex interplay of rights, responsibilities, and relationships that shape our interactions with the natural world.

Through a lens of compassion and empathy, we are called to bear witness to the joys and sorrows, the triumphs and tragedies, of non-human beings, recognizing their capacity for suffering, joy, and emotional connection. From the majestic elephants that roam the savannahs of Africa to the humble chickens that inhabit our backyard coops, each individual possesses a unique constellation of experiences, desires, and needs that demand our attention and respect.

In the exploration of ethical considerations, we are called to embrace a paradigm of moral consideration that extends beyond the confines of speciesism and embraces the intrinsic value of all living beings. Through acts of compassion, advocacy, and stewardship, we can forge a more compassionate and inclusive world—one that honors the inherent worth and dignity of every sentient creature that calls our planet home.

In conclusion, "Animal Consciousness / Bridging Worlds" stands as a testament to the power of empathy, compassion, and ethical reflection in our quest to navigate the depths of animal consciousness. Through a thoughtful exploration of emotional lives, social dynamics, and ethical considerations, Bienuel Tenio invites us to transcend the boundaries of human exceptionalism and embrace a more inclusive vision of consciousness—one that honors the richness and diversity of sentient experience that animates the natural world.

Chapter 5: Language, Communication, and Symbolism

I. Communication Beyond Words: Examining Nonverbal and Symbolic Language.

In the intricate web of conscious experience, language emerges as a powerful tool for communication and expression—a bridge that connects individuals across species, cultures, and contexts. In "Animal Consciousness / Bridging Worlds," Bienuel Tenio invites readers to explore the rich tapestry of nonverbal and symbolic language that animates the natural world, transcending the boundaries of human-centric perspectives to embrace the diversity of communicative modalities that shape conscious awareness.

At its essence, communication extends far beyond the confines of spoken words, encompassing a rich array of gestures, postures, vocalizations, and symbols that convey meaning and intent. From the rhythmic dances of honeybees that signal the location of nectar to the intricate patterns of song that define the social dynamics of bird flocks, animals engage in a dynamic symphony of communication that reflects the depth and complexity of conscious awareness.

Through a lens of symbolic language, we glimpse the rich tapestry of meaning that permeates the natural world—a mosaic of symbols, signs, and signals that speak to the interconnectedness of life. From the ritualized displays of courtship in birds to the elaborate hierarchies of dominance in primate societies, each species possesses its own unique lexicon of communication that reflects the richness and diversity of conscious experience.

In the exploration of nonverbal and symbolic language, we are called to transcend the limitations of human-centric perspectives and embrace the universality of communicative modalities that unite us with the web of life. Through the lens of language, we glimpse the raw beauty and complexity of conscious awareness, forging connections that transcend species boundaries and unite us in a shared tapestry of existence.

II. Case Studies: Extraordinary Examples of Animal Communication.

As we journey deeper into the heart of language and communication, we encounter extraordinary examples of animal communication that challenge our understanding of consciousness and cognition. In "Animal Consciousness / Bridging Worlds," Bienuel Tenio illuminates the profound richness and diversity of communicative modalities that populate the natural world, inviting readers to explore the intricate tapestry of signals, signs, and symbols that shape conscious awareness.

From the haunting calls of whales that reverberate through the depths of the ocean to the intricate dances of bees that convey the location of food sources, animals engage in a rich array of communicative behaviors that reflect the depth and complexity of conscious experience. Through a series of case studies, we encounter the remarkable capacities of non-human beings to convey meaning, intent, and emotion through their vocalizations, gestures, and behaviors.

In the exploration of case studies, we are called to transcend human-centric perspectives and embrace the diversity of communicative modalities that populate the natural world. Through the eyes of animals, we glimpse the raw beauty and vulnerability of conscious awareness, forging connections that transcend species boundaries and unite us in a shared tapestry of existence.

III. Exploring the Limits of Human Understanding: Deciphering Animal Languages.

At the intersection of language and consciousness lies the profound challenge of deciphering the languages of non-human beings—a task that pushes the boundaries of human understanding and perception. In "Animal Consciousness / Bridging Worlds," Bienuel Tenio invites readers to embark on a transformative journey into the depths of animal languages, exploring the intricate tapestry of signals, signs, and symbols that shape conscious awareness.

As we confront the limits of human understanding, we are called to embrace humility and curiosity in our quest to decipher the languages of non-human beings. From the complex vocalizations of dolphins that convey a rich array of meanings to the intricate displays of body language in primates that reflect the dynamics of social interaction, animals engage in a diverse array of communicative behaviors that challenge our assumptions and expand our horizons.

In the exploration of animal languages, we are confronted with the profound complexity and diversity of conscious experience that permeates the natural world. Through the lens of language, we glimpse the raw beauty and vulnerability of sentient beings, forging connections that transcend species boundaries and unite us in a shared tapestry of existence.

In conclusion, "Animal Consciousness / Bridging Worlds" stands as a testament to the power of language, communication, and symbolism in our quest to navigate the depths of conscious awareness. Through a thoughtful exploration of nonverbal and symbolic language, case studies of extraordinary animal communication, and the challenges of deciphering animal languages, Bienuel Tenio invites us to transcend the limitations of human perception and embrace a more inclusive vision of consciousness—one that honors the richness and diversity of sentient experience that animates the natural world.

Chapter 6: Consciousness and Intelligence

I. Problem-Solving Abilities: Insights into Animal Intelligence.

In the intricate tapestry of consciousness, intelligence emerges as a beacon of light—a guiding force that illuminates the path toward understanding the depths of animal minds. In "Animal Consciousness / Bridging Worlds," Bienuel Tenio embarks on a profound exploration of consciousness and intelligence, inviting readers to delve into the rich tapestry of problem-solving abilities that animate the natural world.

At its essence, problem-solving serves as a hallmark of intelligence—a testament to the adaptive capacities and cognitive flexibility of sentient beings. From the intricate puzzles of the natural world to the challenges of navigating complex social landscapes, animals engage in a dynamic symphony of problem-solving behaviors that reflect the depth and complexity of conscious awareness.

Through a lens of problem-solving abilities, we glimpse the raw beauty and complexity of conscious experience—a mosaic of strategies, insights, and innovations that shape the contours of sentient minds. From the ingenious foraging techniques of corvids that defy conventional expectations to the remarkable feats of memory in elephants that span generations, each species possesses its own unique repertoire of problem-solving skills that reflect the richness and diversity of conscious experience.

In the exploration of problem-solving abilities, we are called to transcend human-centric perspectives and embrace the universality of cognitive capacities that unite us with the web of life. Through the lens of intelligence, we glimpse the raw beauty and complexity of conscious awareness, forging connections that transcend species boundaries and unite us in a shared tapestry of existence.

II. Tool Use and Innovation: Demonstrations of Cognitive Flexibility.

Central to the fabric of consciousness lies the remarkable capacity for tool use and innovation—a testament to the cognitive flexibility and ingenuity of sentient beings. In "Animal Consciousness / Bridging Worlds," Bienuel Tenio illuminates the profound role of tool use and innovation in shaping the contours of conscious experience, inviting readers to explore the rich tapestry of cognitive abilities that animate the natural world.

At its essence, tool use represents a hallmark of intelligence—a testament to the adaptive capacities and problem-solving skills of sentient beings. From the intricate tool kits of chimpanzees that enable them to extract termites from mounds to the sophisticated fishing techniques of dolphins that capture prey with precision, animals engage in a dynamic symphony of tool use and innovation that reflects the depth and complexity of conscious awareness.

Through a lens of cognitive flexibility, we glimpse the raw beauty and complexity of conscious experience—a mosaic of behaviors, insights, and innovations that shape the contours of sentient minds. From the playful experimentation of otters that fashion tools from shells to the creative problem-solving of crows that fashion hooks from twigs, each species possesses its own unique repertoire of cognitive abilities that reflect the richness and diversity of conscious experience.

In the exploration of tool use and innovation, we are called to transcend human-centric perspectives and embrace the universality of cognitive capacities that unite us with the web of life. Through the lens of intelligence, we glimpse the raw beauty and complexity of conscious awareness, forging connections that transcend species boundaries and unite us in a shared tapestry of existence.

III. The Enigma of Animal Minds: Challenging Anthropocentric Perspectives.

As we navigate the depths of consciousness and intelligence, we are confronted with the profound enigma of animal minds—a mystery that defies easy explanation and challenges our anthropocentric perspectives. In "Animal Consciousness / Bridging Worlds," Bienuel Tenio invites readers to confront the complexity and diversity of animal minds, exploring the intricate tapestry of cognitive abilities that animate the natural world.

At its essence, the enigma of animal minds reflects the limitations of human understanding—a recognition that conscious experience extends far beyond the confines of human perception and cognition. From the intricate problem-solving abilities of octopuses that rival those of primates to the remarkable tool use of New Caledonian crows that rivals human innovation, animals engage in a rich array of cognitive behaviors that challenge our assumptions and expand our horizons.

Through a lens of humility and curiosity, we are called to embrace the mystery and wonder of animal minds, recognizing the inherent value and dignity of all sentient beings. From the playful antics of dolphins that echo our own capacity for joy to the mournful cries of elephants that reflect our shared capacity for grief, each species possesses its own unique constellation of cognitive abilities that reflect the richness and diversity of conscious experience.

In the exploration of animal minds, we are called to transcend the limitations of human perception and embrace a more inclusive vision of consciousness—one that honors the richness and diversity of sentient experience that animates the natural world. Through acts of compassion, empathy, and stewardship, we can forge a more compassionate and inclusive world—one that celebrates the inherent worth and dignity of every sentient creature that calls our planet home.

In conclusion, "Animal Consciousness / Bridging Worlds" stands as a testament to the power of intelligence, problem-solving abilities, and cognitive flexibility in our quest to navigate the depths of conscious awareness. Through a thoughtful exploration of animal minds, Bienuel Tenio invites us to transcend the boundaries of human exceptionalism and embrace a more inclusive vision of consciousness—one that honors the richness and diversity of sentient experience that animates the natural world.

Chapter 7: Consciousness in the Wild: Observations and Discoveries

In the vast expanse of the natural world, consciousness unfurls in a myriad of forms, inviting us to embark on a journey of discovery and revelation. In "Animal Consciousness / Bridging Worlds," authored by Bienuel Tenio, the exploration of consciousness extends beyond the confines of human-centric perspectives, beckoning us to delve into the depths of animal consciousness as it manifests in the wild. Through field studies, insights from captive settings, and revelations from unexpected sources, we are invited to contemplate the rich tapestry of conscious awareness that animates the natural world.

I. Field Studies: Observing Animal Behavior in Natural Habitats.

Field studies represent a window into the soul of the natural world—a canvas upon which the intricate patterns of conscious awareness unfold in breathtaking detail. In "Animal Consciousness / Bridging Worlds," Bienuel Tenio illuminates the profound insights gleaned from observing animal behavior in their natural habitats, inviting readers to bear witness to the raw beauty and complexity of conscious experience.

In the untamed wilderness, animals navigate a world shaped by the rhythms of nature, forging connections, and engaging in behaviors that reflect the richness of their inner lives. From the majestic elephants that traverse the savannahs of Africa to the elusive octopuses that roam the depths of the ocean, each species offers a unique glimpse into the mysteries of consciousness, inviting us to explore the depths of sentient experience.

Through the lens of field studies, we are called to transcend human-centric perspectives and embrace the diversity of conscious awareness that permeates the natural world. From the subtle nuances of social interactions to the intricate dynamics of predator-prey relationships, every encounter offers a window into the rich tapestry of life, forging connections that transcend species boundaries and unite us in a shared journey of exploration and discovery.

II. Insights from Captive Settings: Ethical Considerations and Research Challenges.

In the realm of captive settings, ethical considerations and research challenges cast a shadow over our quest to unravel the mysteries of animal consciousness. In "Animal Consciousness / Bridging Worlds," Bienuel Tenio navigates the complex terrain of captivity, inviting readers to confront the ethical dilemmas and methodological limitations that shape our understanding of conscious awareness.

In the controlled environment of captivity, animals are subject to a myriad of stressors and constraints that can impact their behavior and well-being. From the confines of cages to the artificial stimuli of laboratory settings, captive animals face a host of challenges that can distort our perception of their true nature and cognitive capacities.

Despite these limitations, captive settings offer valuable insights into the intricacies of conscious awareness, providing researchers with a controlled environment in which to explore the depths of animal cognition. From the pioneering studies of primates in laboratory settings to the innovative techniques used to assess cognitive abilities in dolphins and elephants, captive research has yielded a wealth of knowledge that enriches our understanding of conscious experience.

In the exploration of captive settings, we are called to confront the ethical considerations and research challenges that shape our quest for knowledge. By embracing compassion, empathy, and respect for the inherent dignity of all living beings, we can navigate the complexities of captivity with integrity and humility, honoring the intrinsic value of every sentient creature that shares our planet.

III. Revelations from Unexpected Sources: Surprising Discoveries in Animal Consciousness.

In the labyrinth of existence, consciousness reveals itself in unexpected places, inviting us to expand our horizons and challenge our preconceived notions. In "Animal Consciousness / Bridging Worlds," Bienuel Tenio illuminates the surprising discoveries and unexpected revelations that emerge from the depths of animal consciousness, inviting readers to embrace the mystery and wonder of sentient experience.

From the humble insects that inhabit our gardens to the enigmatic creatures that dwell in the depths of the ocean, every corner of the natural world offers a glimpse into the infinite tapestry of conscious awareness. Through a series of surprising discoveries, we encounter the remarkable capacities of non-human beings to navigate the complexities of life, challenging our assumptions and expanding our understanding of consciousness.

In the exploration of unexpected sources, we are called to embrace humility and curiosity in our quest for knowledge. By remaining open to the wonders of the natural world and the mysteries that lie beyond our comprehension, we can cultivate a deeper appreciation for the richness and diversity of conscious experience, forging connections that transcend the boundaries of species and unite us in a shared journey of exploration and discovery.

In conclusion, "Animal Consciousness / Bridging Worlds" stands as a testament to the power of observation, inquiry, and reverence in our quest to unravel the mysteries of consciousness. Through field studies, insights from captive settings, and revelations from unexpected sources, Bienuel Tenio invites us to embrace the wonder and awe of sentient experience, forging connections that unite us with the vibrant tapestry of life.

Chapter 8: The Human-Animal Connection

In the intricate tapestry of existence, the bond between humans and animals transcends the boundaries of species, inviting us to explore the depths of empathy, understanding, and interconnectedness. Authored by Bienuel Tenio, "Animal Consciousness / Bridging Worlds" delves into the profound connection between humans and animals, offering insights into the rich tapestry of conscious awareness that binds us together. Through observations and discoveries, we uncover the intricacies of companion animals, historical perspectives, and shared experiences that shape our understanding of the human-animal connection.

I. Companion Animals: Bonds, Empathy, and Understanding.

Companion animals occupy a unique place in the human experience, serving as steadfast companions, sources of comfort, and mirrors of our own emotions and behaviors. In "Animal Consciousness / Bridging Worlds," Bienuel Tenio explores the profound bonds that form between humans and their animal companions, illuminating the depths of empathy, understanding, and mutual affection that define the human-animal connection.

At the heart of the human-animal bond lies a profound reciprocity—a dynamic interplay of care, companionship, and shared experiences that enriches the lives of both humans and animals. From the loyal dogs that greet us with wagging tails and eager eyes to the gentle purrs of contentment that emanate from our feline friends, companion animals offer a source of solace, joy, and unconditional love in an often chaotic and uncertain world.

Through the lens of empathy, we glimpse the depths of emotional intelligence that animate the human-animal bond, forging connections that transcend language, culture, and species boundaries. From the profound grief that accompanies the loss of a beloved pet to the jubilant celebrations of milestones and triumphs, our lives are intertwined with those of our animal companions in ways that defy explanation and enrich the fabric of our existence.

II. Historical Perspectives: Cultural Views on Animal Consciousness.

Throughout history, cultures around the world have grappled with questions of animal consciousness, offering a kaleidoscope of perspectives that reflect the diversity of human experience and belief. In "Animal Consciousness / Bridging Worlds," Bienuel Tenio traces the evolution of cultural views on animal consciousness, illuminating the myriad ways in which humans have sought to understand and interpret the minds of other beings.

From ancient civilizations that revered animals as sacred guardians of the natural world to modern societies that grapple with questions of ethics and animal welfare, our views on animal consciousness have evolved in tandem with our understanding of the natural world and our place within it. Through myths, legends, and religious texts, we encounter a tapestry of beliefs and traditions that speak to the enduring connection between humans and animals—a connection that transcends time, space, and cultural boundaries.

In the exploration of historical perspectives, we are called to confront the complexities of human-animal relationships and the ethical dilemmas that accompany our interactions with other beings. By embracing humility, compassion, and cultural sensitivity, we can navigate the rich tapestry of beliefs and traditions that shape our understanding of animal consciousness, forging connections that unite us with the vibrant diversity of human experience.

III. Shared Experiences: Learning from Animals and Enriching Human Lives.

In the shared tapestry of human-animal relationships, we discover a wealth of wisdom, insight, and inspiration that enriches our lives and deepens our understanding of the world around us. In "Animal Consciousness / Bridging Worlds," Bienuel Tenio invites readers to explore the transformative power of shared experiences, illuminating the profound lessons that animals impart and the ways in which they shape our understanding of conscious awareness.

From the humble wisdom of farm animals that teach us the value of hard work and perseverance to the playful antics of dolphins that remind us of the joy and wonder of life, animals offer a mirror through which we glimpse the depths of our own humanity. Through shared experiences of compassion, empathy, and understanding, we forge connections that transcend the boundaries of species and unite us in a shared journey of growth, discovery, and transformation.

In the exploration of shared experiences, we are called to embrace the lessons that animals teach us and the profound impact they have on our lives. By cultivating reverence, gratitude, and respect for all living beings, we can nurture a deeper appreciation for the interconnectedness of life and the beauty of conscious awareness that animates the natural world.

In conclusion, "Animal Consciousness / Bridging Worlds" stands as a testament to the power of the human-animal connection in shaping our understanding of conscious awareness and enriching the fabric of our existence. Through companion animals, historical perspectives, and shared experiences, Bienuel Tenio invites us to embrace the depths of empathy, understanding, and interconnectedness that define the human-animal bond, forging connections that unite us with the vibrant tapestry of life.

Chapter 9: Consciousness and Conservation

In the intricate tapestry of life, the conservation of biodiversity and the preservation of ecosystems stand as paramount concerns for humanity. Authored by Bienuel Tenio, "Animal Consciousness / Bridging Worlds" explores the profound intersection of consciousness and conservation, inviting readers to contemplate the intricate relationships between sentient beings and the natural world. Through observations and discoveries, we uncover the complexities of biodiversity, conservation ethics, and the integration of consciousness studies into conservation efforts.

I. Biodiversity and Ecosystem Health: Recognizing Animals as Sentient Beings.

At the heart of conservation lies a fundamental recognition of the intrinsic value and interconnectedness of all living beings. In "Animal Consciousness / Bridging Worlds," Bienuel Tenio illuminates the vital role that animals play in maintaining the health and resilience of ecosystems, inviting readers to recognize animals as sentient beings whose welfare is intricately linked to the well-being of the planet.

Biodiversity serves as a cornerstone of ecosystem health, encompassing the rich tapestry of life that sustains the web of existence. From the bustling communities of insects that pollinate our crops to the majestic predators that regulate populations and maintain ecological balance, animals contribute to the resilience and vitality of ecosystems in ways both seen and unseen.

Through the lens of consciousness, we glimpse the profound interconnectedness of all living beings—a web of relationships that transcends species boundaries and unites us in a shared tapestry of existence. By recognizing animals as sentient beings with their own intrinsic value and dignity, we can cultivate a deeper appreciation for the role they play in shaping the fabric of life and the ecosystems upon which we all depend.

II. Conservation Ethics: Protecting Animal Consciousness in the Wild.

As stewards of the natural world, we bear a profound responsibility to protect and preserve the diversity of life that surrounds us. In "Animal Consciousness / Bridging Worlds," Bienuel Tenio navigates the complex terrain of conservation ethics, inviting readers to confront the ethical dilemmas and moral imperatives that shape our interactions with other beings.

Conservation ethics demand a recognition of the rights and interests of animals, as well as a commitment to uphold principles of compassion, respect, and stewardship in our interactions with the natural world. From the ethical imperative to mitigate human impacts on wildlife populations to the moral obligation to safeguard the habitats upon which they depend, conservation ethics call upon us to act with integrity and humility in our quest to protect animal consciousness in the wild.

Through the lens of consciousness, we are called to confront the inherent worth and dignity of all living beings, recognizing their capacity for suffering, joy, and emotional connection. By embracing compassion, empathy, and reverence for the web of life, we can forge a more harmonious and sustainable relationship with the natural world, one that honors the interconnectedness of all living beings and safeguards the diversity of life for generations to come.

III. Future Directions: Integrating Consciousness Studies into Conservation Efforts.

As we confront the challenges of the Anthropocene era, the integration of consciousness studies into conservation efforts emerges as a vital frontier in our quest to protect and preserve the diversity of life. In "Animal Consciousness / Bridging Worlds," Bienuel Tenio explores the potential of consciousness studies to inform and enrich conservation practices, inviting readers to embrace a holistic approach to biodiversity conservation that honors the intrinsic value of all living beings.

By integrating consciousness studies into conservation efforts, we can deepen our understanding of the complex relationships between sentient beings and their environments, illuminating the ways in which consciousness shapes behaviors, relationships, and ecological dynamics. From the study of animal cognition and emotion to the exploration of interspecies communication and social dynamics, consciousness studies offer valuable insights that can inform and guide conservation strategies in an increasingly interconnected world.

In the pursuit of conservation, we are called to transcend disciplinary boundaries and embrace a holistic approach that honors the richness and diversity of conscious awareness that permeates the natural world. By integrating consciousness studies into conservation efforts, we can forge a more compassionate, inclusive, and effective approach to biodiversity conservation—one that recognizes the intrinsic value of all living beings and safeguards the diversity of life for future generations.

In conclusion, "Animal Consciousness / Bridging Worlds" stands as a testament to the profound intersection of consciousness and conservation in our quest to protect and preserve the diversity of life on Earth. Through observations and discoveries, Bienuel Tenio invites us to embrace a holistic approach to biodiversity conservation that honors the intrinsic value of all living beings and safeguards the interconnected web of life upon which we all depend.

Chapter 10: Beyond Boundaries: Philosophical and Metaphysical Reflections

In the culmination of "Animal Consciousness / Bridging Worlds," authored by Bienuel Tenio, we embark on a journey beyond the confines of empirical inquiry and scientific exploration. In this chapter, we delve into the philosophical and metaphysical dimensions of animal consciousness, inviting readers to contemplate the mysteries, paradoxes, and interconnectedness that shape our understanding of conscious awareness.

I. The Nature of Consciousness: Exploring Mysteries and Paradoxes.

At the heart of our inquiry lies the enigma of consciousness—an elusive and ineffable phenomenon that defies easy definition and comprehension. In "Animal Consciousness / Bridging Worlds," Bienuel Tenio invites readers to explore the nature of consciousness, probing the depths of mysteries and paradoxes that lurk beneath the surface of human understanding.

Consciousness, in its essence, remains shrouded in mystery—a riddle wrapped in an enigma that has captivated the minds of philosophers, scientists, and mystics throughout the ages. From the subjective experiences of sentient beings to the intricate web of neural processes that underlie conscious awareness, the nature of consciousness eludes our grasp, inviting us to embark on a journey of exploration and discovery.

As we confront the mysteries of consciousness, we encounter paradoxes that challenge our assumptions and expand our horizons. From the dualities of subjectivity and objectivity to the conundrums of free will and determinism, consciousness invites us to embrace the complexity and ambiguity of existence, transcending the limitations of linear thinking and embracing the richness of conscious experience.

II. Interconnectedness and Interdependence: Rethinking Humanity's Relationship with Animals.

In the tapestry of existence, humanity is but one thread woven into the fabric of life—a humble participant in a grand symphony of interconnectedness and interdependence. In "Animal Consciousness / Bridging Worlds," Bienuel Tenio invites readers to rethink humanity's relationship with animals, illuminating the profound web of connections that unite us with the natural world.

At its essence, the interconnectedness of all living beings lies at the heart of conscious awareness—a recognition of the fundamental unity and shared destiny that binds us together. From the smallest insect to the mightiest whale, every creature plays a vital role in the intricate web of life, contributing to the diversity, resilience, and beauty of the natural world.

As stewards of the Earth, we are called to embrace a paradigm of compassion, respect, and reverence for all living beings, recognizing the inherent worth and dignity of every sentient creature that shares our planet. Through acts of kindness, empathy, and understanding, we can nurture a deeper appreciation for the interconnectedness of life and cultivate a more harmonious relationship with the natural world.

III. Towards a Holistic Understanding: Integrating Science, Ethics, and Spirituality.

In our quest for understanding, we are called to transcend the boundaries of disciplinary silos and embrace a holistic vision of consciousness—one that integrates science, ethics, and spirituality into a unified framework of inquiry. In "Animal Consciousness / Bridging Worlds," Bienuel Tenio calls upon readers to embrace a holistic understanding of conscious awareness, recognizing the interconnectedness of all living beings and the profound mysteries that unite us with the cosmos.

Through the lens of science, we gain insights into the workings of the physical universe, unraveling the mysteries of the cosmos and probing the depths of conscious experience. Through the lens of ethics, we confront the moral imperatives that shape our interactions with the natural world, embracing compassion, empathy, and reverence for all living beings. And through the lens of spirituality, we cultivate a sense of awe, wonder, and reverence for the profound mysteries of existence, embracing the interconnectedness of all life and the boundless potential of conscious awareness.

In the integration of science, ethics, and spirituality, we discover a tapestry of meaning and purpose that transcends the limitations of human understanding, inviting us to embark on a journey of exploration and discovery that spans the depths of the cosmos and the reaches of the human soul.

In conclusion, "Animal Consciousness / Bridging Worlds" stands as a testament to the power of inquiry, exploration, and contemplation in our quest to unravel the mysteries of conscious awareness. Through philosophical and metaphysical reflections, Bienuel Tenio invites us to embrace the mysteries, paradoxes, and interconnectedness that shape our understanding of conscious experience, forging connections that unite us with the vibrant tapestry of life.

Chapter 11: Conclusion

In the concluding chapter of "Animal Consciousness / Bridging Worlds: Exploring the Depths of Animal Consciousness" by Bienuel Tenio, we embark on a journey of reflection and contemplation, weaving together the threads of insight and discovery that have unfolded throughout our exploration of animal consciousness. As we navigate the depths of philosophical and metaphysical inquiry, we are called to summarize key insights and findings, look ahead to future directions in research, and embrace a world where all conscious beings matter.

I. Summarizing Key Insights and Findings.

Our journey into the depths of animal consciousness has revealed a tapestry of insight and revelation—a mosaic of experiences, behaviors, and relationships that shape the contours of conscious awareness. Through the lens of scientific inquiry and philosophical reflection, we have glimpsed the rich tapestry of sentient experience that animates the natural world, transcending the boundaries of human-centric perspectives to embrace the diversity and complexity of conscious awareness.

At its essence, animal consciousness is a testament to the richness and diversity of life—a celebration of the interconnectedness and interdependence that bind us together in a shared tapestry of existence. From the profound emotional lives of companion animals to the intricate social dynamics of wild populations, animals offer a mirror through which we glimpse the raw beauty and vulnerability of conscious awareness, forging connections that transcend species boundaries and unite us in a shared journey of exploration and discovery.

As we reflect on key insights and findings, we are called to embrace humility and curiosity in our quest for understanding, recognizing the limitations of human perception and the boundless mysteries that lie beyond our grasp. Through acts of compassion, empathy, and reverence for all living beings, we can nurture a deeper appreciation for the richness and diversity of conscious experience that permeates the natural world, fostering a more inclusive and compassionate vision of existence.

II. Looking Ahead: Future Directions in Animal Consciousness Research.

As we stand at the threshold of the unknown, we are called to look ahead to future directions in animal consciousness research, embracing the challenges and opportunities that lie on the horizon. In the ever-evolving landscape of scientific inquiry, new technologies, methodologies, and interdisciplinary collaborations offer unprecedented avenues for exploration and discovery, inviting us to push the boundaries of knowledge and understanding.

From the frontiers of neuroscience and cognitive psychology to the depths of ecological and ethological research, future investigations hold the promise of uncovering new insights into the nature of conscious awareness and the complexities of sentient experience. By embracing a holistic and integrative approach to research, we can transcend the limitations of disciplinary boundaries and forge new connections that illuminate the mysteries of consciousness from myriad perspectives.

In the pursuit of future directions, we are called to embrace a spirit of curiosity, wonder, and collaboration, recognizing that the quest for knowledge is a collective endeavor that transcends individual ambitions and agendas. Through shared exploration and discovery, we can unlock the secrets of consciousness and unravel the mysteries of existence, forging a path toward a more enlightened and compassionate world.

III. Call to Action: Embracing a World Where All Conscious Beings Matter.

As we reflect on our journey into the depths of animal consciousness, we are called to action—to embrace a world where all conscious beings matter, and where compassion, empathy, and reverence guide our interactions with the natural world. In a world marked by ecological crisis and biodiversity loss, the imperative to protect and preserve the diversity of life has never been more urgent.

Through acts of stewardship, advocacy, and ethical engagement, we can champion the rights and welfare of all living beings, ensuring that every sentient creature is afforded the dignity, respect, and compassion it deserves. From the humblest insect to the mightiest whale, every creature plays a vital role in the intricate web of life, contributing to the richness, resilience, and beauty of the natural world.

In the face of adversity, we are called to stand as guardians of the Earth, protecting and preserving the precious tapestry of life for future generations. By embracing a spirit of reverence and awe for the wonders of the natural world, we can forge a path toward a more harmonious and sustainable future—a future where all conscious beings are valued, cherished, and celebrated as integral members of the global community.

In conclusion, "Animal Consciousness / Bridging Worlds" stands as a testament to the power of inquiry, exploration, and reflection in our quest to unravel the mysteries of conscious awareness. Through philosophical and metaphysical reflections, we have glimpsed the richness and diversity of sentient experience that animates the natural world, forging connections that unite us with the vibrant tapestry of life. As we look ahead to future directions and embrace a world where all conscious beings matter, let us embark on a journey of compassion, empathy, and reverence, guided by the wisdom and insight that arise from our exploration of the depths of animal consciousness.

www.ingramcontent.com/pod-product-compliance
Lightning Source LLC
Chambersburg PA
CBHW070420230526
45471CB00006B/2895